Cram101 Textbook Outlines to accompany:

Signals, Systems, and Transforms

Charles L Phillips, 4th Edition

A Content Technologies Inc. publication (c) 2012.

Learning System

Cram101 Textbook Outlines is a learning system. The notes in this book are the highlights of your textbook, you will never have to highlight a book again.

How to use this book. Take this book to class, it is your notebook for the lecture. The notes and highlights on the left hand side of the pages follow the outline and order of the textbook. All you have to do is follow along while your instructor presents the lecture. Circle the items emphasized in class and add other important information on the right side. With Cram101 Textbook Outlines you'll spend less time writing and more time listening. Learning becomes more efficient.

Cram101.com Online

Increase your studying efficiency by using Cram101.com's practice tests and online reference material. It is the perfect complement to Cram101 Textbook Outlines. Use self-teaching matching tests or simulate in-class testing with comprehensive multiple choice tests, or simply use Cram's true and false tests for quick review. Cram101.com even allows you to enter your in-class notes for an integrated studying format combining the textbook notes with your class notes.

Visit **www.Cram101.com**, click Sign Up at the top of the screen, and enter **DK73DW11860** in the promo code box on the registration screen. Your access to www.Cram101.com is discounted by 50% because you have purchased this book. Sign up and stop highlighting textbooks forever.

 ISBN(s): 9781618308825. PUBE-8.2011716

Signals, Systems, and Transforms
Charles L Phillips, 4th

CONTENTS

1. INTRODUCTION 2
2. CONTINUOUS-TIME SIGNALS AND SYSTEMS 16
3. CONTINUOUS-TIME LINEAR TIME-INVARIANT SYSTEMS 30
4. FOURIER SERIES 42
5. THE FOURIER TRANSFORM 54
6. APPLICATIONS OF THE FOURIER TRANSFORM 70
7. THE LAPLACE TRANSFORM 84
8. STATE VARIABLES FOR CONTINUOUS-TIME SYSTEMS 96
9. DISCRETE-TIME SIGNALS AND SYSTEMS 104
10. DISCRETE-TIME LINEAR TIME-INVARIANT SYSTEMS 114
11. THE z-TRANSFORM 124

Chapter 1. INTRODUCTION

Signal	· Signal (information theory), a physical quantity that can carry information. · Signal processing, the field of techniques used to extract information from Signals · Cell Signaling, in biology, the system of communication that governs basic cellular activities and coordinates cell actions · Signal (physiology), an electrochemical communication activity in an organism · Two theories of `honest Signalling` in different disciplines: · Signalling theory in evolutionary biology, a theory stating how organisms Signal their state to others with respect to evolutionary survival and reproductive fitness · Signalling (economics), how market actors can Signal their intentions and values to others in economics · Signalling game, Signalling theories in evolutionary biology and economics expressed in terms of game theory · Signal (computing), an event, message, or data structure transmitted between computational processes · Signal programming, a variety of event-driven programming · Signals and slots, a Signal-driven software design pattern · Signals used in various kinds of transport: · Traffic light · Railway Signal · Semaphore

· Beacon

· International maritime Signal flags, using the International Code of Signals

· Signals (military), a name that refers to military communications

· Distress Signal

· Smoke Signal

· Signalling (telecommunications), a part of some communication protocols

· Recognition Signal in human, technical or biological communications

· Signal (band), a Bulgarian rock band

· Signal (American band), an American AOR band

· Signals (album), an album by progressive rock band Rush

· Signal (magazine), a Nazi propaganda magazine published in occupied Europe during World War II

· The Signal, a 2007 horror film

· Signal (toothpaste)

· Signal 1 and Signal 2 radio stations

· Signaal (now Thales Nederland), a Netherlands based defence company

· Signal Systems, a telecommunications company of the McLean Group of Companies

· Signal - The Southeast Electronic Music Festival, an annual music festival held in the United States

· USS Signal, the name of more than one United States Navy ship

· Any implicit, subtle, or disguised message, in situations such as:

· Partnership card games)

· Signals in legal citations

· Kent (game) a card game also known as Signal or Kemp

·

System

In computer science and information science, System could also be a method or an algorithm. Again, an example will illustrate: There are Systems of counting, as with Roman numerals, and various Systems for filing papers, or catalogues, and various library Systems, of which the Dewey Decimal System is an example. This still fits with the definition of components which are connected together (in this case in order to facilitate the flow of information).

Analogue

The analogue switch is an electronic component that behaves in a similar way to a relay, but has no moving parts. The switching element is normally a MOSFET, which is a type of transistor. The control input to the switch is a standard CMOS or TTL logic input, which is shifted by internal circuitry to a suitable voltage for switching the MOSFET. The result is that a logic 0 on the control input causes the MOSFET to have a high resistance, so that the switch is off, and a logic 1 on the input causes the MOSFET to have a low resistance, so that the switch is on.

Equation

An equation is a mathematical statement that asserts the equality of two expressions. equations consist of the expressions that are to be equal on opposite sides of an equal sign, as in

$$x + 3 = 5$$

$$9 - y = 7$$

One use of equations is in mathematical identities, assertions that are true independent of the values of any variables contained within them. For example, for any given value of x it is true that

$$x(x-1) = x^2 - x\,.$$

However, equations can also be correct for only certain values of the variables.

Analog signal

An Analog or analogue signal is any continuous signal for which the time varying feature (variable) of the signal is a representation of some other time varying quantity, i.e analogous to another time varying signal. It differs from a digital signal in terms of small fluctuations in the signal which are meaningful. Analog is usually thought of in an electrical context; however, mechanical, pneumatic, hydraulic, and other systems may also convey Analog signals.

Digital

A Digital system is a data technology that uses discrete (discontinuous) values. By contrast, non-Digital (or analog) systems use a continuous range of values to represent information. Although Digital representations are discrete, the information represented can be either discrete, such as numbers, letters or icons, or continuous, such as sounds, images, and other measurements of continuous systems.

Digital signal

The term Digital signal is used to refer to more than one concept. It can refer to discrete-time signals that have a discrete number of levels, for example a sampled and quantified analog signal, or to the continuous-time waveform signals in a digital system, representing a bit-stream. In the first case, a signal that is generated by means of a digital modulation method which is considered as converted to an analog signal, while it is considered as a Digital signal in the second case.

Digital signal processing

Digital signal processing is concerned with the representation of the signals by a sequence of numbers or symbols and the processing of these signals. Digital signal processing and analog signal processing are subfields of signal processing. Digital signal processing includes subfields like: audio and speech signal processing, sonar and radar signal processing, sensor array processing, spectral estimation, statistical signal processing, digital image processing, signal processing for communications, biomedical signal processing, seismic data processing, etc.

Discrete time

Discrete time is non-continuous time. Sampling at non-continuous times results in discrete-time samples. For example, a newspaper may report the price of crude oil once every 24 hours.

Linear differential equation

In mathematics, a Linear differential equation is of the form

$$Ly = f$$

where the differential operator L is a linear operator, y is the unknown function (such as a function of time y(t)), and the right hand side f is a given function of the same nature as y (called the source term). For a function dependent on time we may write the equation more expressively as

$$Ly(t) = f(t)$$

and, even more precisely by bracketing

$$L[y(t)] = f(t)$$

The linear operator L may be considered to be of the form

$$L_n(y) \equiv \frac{d^n y}{dt^n} + A_1(t)\frac{d^{n-1}y}{dt^{n-1}} + \cdots + A_{n-1}(t)\frac{dy}{dt} + A_n(t)y$$

The linearity condition on L rules out operations such as taking the square of the derivative of y; but permits, for example, taking the second derivative of y. It is convenient to rewrite this equation in an operator form

$$L_n(y) \equiv \left[\,D^n + A_1(t)D^{n-1} + \cdots + A_{n-1}(t)D + A_n(t)\right] y$$

where D is the differential operator d/dt (i.e. Dy = y\` , $D^2y$ = y\`,...

Operational Amplifier

An Operational amplifier, which is often called an op-amp, is a DC-coupled high-gain electronic voltage amplifier with differential inputs and, usually, a single output. Typically the output of the op-amp is controlled either by negative feedback, which largely determines the magnitude of its output voltage gain, or by positive feedback, which facilitates regenerative gain and oscillation. High input impedance at the input terminals (ideally infinite) and low output impedance (ideally zero) are important typical characteristics.

Voltage

Voltage is commonly used as a short name for electrical potential difference. Its corresponding SI unit is the volt (symbol: V, not italicized). Electric potential is a hypothetically measurable physical dimension, and is denoted by the algebraic variable V (italicized).

Analog-to-digital converter

An Analog-to-digital converter is a device which converts continuous signals to discrete digital numbers. The reverse operation is performed by a digital-to-analog converter (DAnalog-to-digital converter).
Typically, an ADC is an electronic device that converts an input analog voltage (or current) to a digital number proportional to the magnitude of the voltage or current.

Digital-to-analog Converter

In electronics, a Digital-to-analog converter (DAC or D-to-A) is a device for converting a digital (usually binary) code to an analog signal (current, voltage or electric charge).
An analog-to-digital converter (ADigital-to-analog converter) performs the reverse operation.

Ideally sampled signal.

Pixel

In digital imaging, a PixelCite error: Closing </ref> missing for <ref> tag. and texel.
The word \`Pixel\` was first published in 1965 by Frederic C. Billingsley of JPL (in Pasadena, CA), to describe the picture elements of video images from space probes to the Moon and Mars.

Time-division multiplexing

Time-division multiplexing (TDM) is a type of digital or (rarely) analog multiplexing in which two or more signals or bit streams are transferred apparently simultaneously as sub-channels in one communication channel, but are physically taking turns on the channel. The time domain is divided into several recurrent timeslots of fixed length, one for each sub-channel. A sample byte or data block of sub-channel 1 is transmitted during timeslot 1, sub-channel 2 during timeslot 2, etc.

Chapter 1. INTRODUCTION

Pulse	In medicine, a person`s Pulse is the arterial palpation of a heartbeat. It can be palpated in any place that allows for an artery to be compressed against a bone, such as at the neck (carotid artery), at the wrist (radial artery), behind the knee (popliteal artery), on the inside of the elbow (brachial artery), and near the ankle joint (posterior tibial artery). The Pulse rate can also be measured by measuring the heart beats directly (the apical Pulse).
Telephone	The Telephone is a telecommunications device that transmits and receives sound, most commonly the human voice. It is one of the most common household appliances in the developed world, and has long been considered indispensable to business, industry and government. The word `Telephone` has been adapted to many languages and is widely recognized around the world.
Time constant	In physics and engineering, the Time constant, usually denoted by the Greek letter τ , is the risetime characterizing the response to a time-varying input of a first-order, linear time-invariant (LTI) system. (That is, a system that can be modeled by a single first order differential equation in time. Examples include the simplest single-stage electrical RC circuits and RL circuits).
Delta modulation	Delta modulation (Delta modulation or Δ-modulation) is an analog-to-digital and digital-to-analog signal conversion technique used for transmission of voice information where quality is not of primary importance. Delta modulation is the simplest form of differential pulse-code modulation (DPCM) where the difference between successive samples is encoded into n-bit data streams. In Delta modulation, the transmitted data is reduced to a 1-bit data stream.

Analogue	The analogue switch is an electronic component that behaves in a similar way to a relay, but has no moving parts. The switching element is normally a MOSFET, which is a type of transistor. The control input to the switch is a standard CMOS or TTL logic input, which is shifted by internal circuitry to a suitable voltage for switching the MOSFET. The result is that a logic 0 on the control input causes the MOSFET to have a high resistance, so that the switch is off, and a logic 1 on the input causes the MOSFET to have a low resistance, so that the switch is on.
Analog signal	An Analog or analogue signal is any continuous signal for which the time varying feature (variable) of the signal is a representation of some other time varying quantity, i.e analogous to another time varying signal. It differs from a digital signal in terms of small fluctuations in the signal which are meaningful. Analog is usually thought of in an electrical context; however, mechanical, pneumatic, hydraulic, and other systems may also convey Analog signals.
Signal	· Signal (information theory), a physical quantity that can carry information. · Signal processing, the field of techniques used to extract information from Signals · Cell Signaling, in biology, the system of communication that governs basic cellular activities and coordinates cell actions · Signal (physiology), an electrochemical communication activity in an organism · Two theories of `honest Signalling` in different disciplines: · Signalling theory in evolutionary biology, a theory stating how organisms Signal their state to others with respect to evolutionary survival and reproductive fitness · Signalling (economics), how market actors can Signal their intentions and values to others in economics · Signalling game, Signalling theories in evolutionary biology and economics expressed in terms of game theory · Signal (computing), an event, message, or data structure transmitted between computational processes

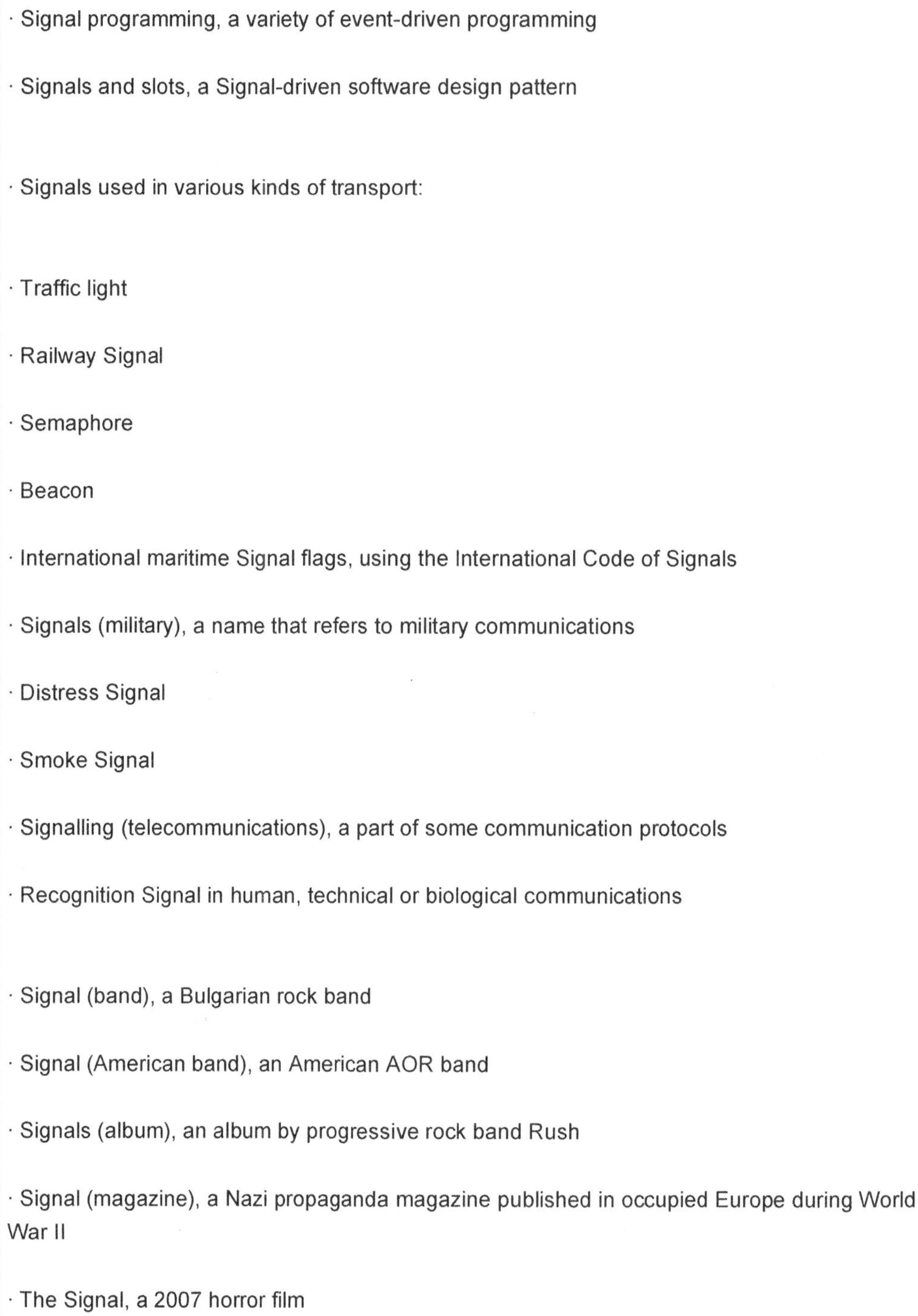

· Signal programming, a variety of event-driven programming

· Signals and slots, a Signal-driven software design pattern

· Signals used in various kinds of transport:

· Traffic light

· Railway Signal

· Semaphore

· Beacon

· International maritime Signal flags, using the International Code of Signals

· Signals (military), a name that refers to military communications

· Distress Signal

· Smoke Signal

· Signalling (telecommunications), a part of some communication protocols

· Recognition Signal in human, technical or biological communications

· Signal (band), a Bulgarian rock band

· Signal (American band), an American AOR band

· Signals (album), an album by progressive rock band Rush

· Signal (magazine), a Nazi propaganda magazine published in occupied Europe during World War II

· The Signal, a 2007 horror film

· Signal (toothpaste)

· Signal 1 and Signal 2 radio stations

· Signaal (now Thales Nederland), a Netherlands based defence company

· Signal Systems, a telecommunications company of the McLean Group of Companies

· Signal - The Southeast Electronic Music Festival, an annual music festival held in the United States

· USS Signal, the name of more than one United States Navy ship

· Any implicit, subtle, or disguised message, in situations such as:

· Partnership card games)

· Signals in legal citations

· Kent (game) a card game also known as Signal or Kemp

·

System	In computer science and information science, System could also be a method or an algorithm. Again, an example will illustrate: There are Systems of counting, as with Roman numerals, and various Systems for filing papers, or catalogues, and various library Systems, of which the Dewey Decimal System is an example. This still fits with the definition of components which are connected together (in this case in order to facilitate the flow of information).
Digital	A Digital system is a data technology that uses discrete (discontinuous) values. By contrast, non-Digital (or analog) systems use a continuous range of values to represent information. Although Digital representations are discrete, the information represented can be either discrete, such as numbers, letters or icons, or continuous, such as sounds, images, and other measurements of continuous systems.

Digital signal	The term Digital signal is used to refer to more than one concept. It can refer to discrete-time signals that have a discrete number of levels, for example a sampled and quantified analog signal, or to the continuous-time waveform signals in a digital system, representing a bit-stream. In the first case, a signal that is generated by means of a digital modulation method which is considered as converted to an analog signal, while it is considered as a Digital signal in the second case.
Digital signal processing	Digital signal processing is concerned with the representation of the signals by a sequence of numbers or symbols and the processing of these signals. Digital signal processing and analog signal processing are subfields of signal processing. Digital signal processing includes subfields like: audio and speech signal processing, sonar and radar signal processing, sensor array processing, spectral estimation, statistical signal processing, digital image processing, signal processing for communications, biomedical signal processing, seismic data processing, etc.
Digital-to-analog Converter	In electronics, a Digital-to-analog converter (DAC or D-to-A) is a device for converting a digital (usually binary) code to an analog signal (current, voltage or electric charge). An analog-to-digital converter (ADigital-to-analog converter) performs the reverse operation. Ideally sampled signal.
Discrete time	Discrete time is non-continuous time. Sampling at non-continuous times results in discrete-time samples. For example, a newspaper may report the price of crude oil once every 24 hours.
Cathode	A Cathode is an electrode through which electric current flows out of a polarized electrical device. Mnemonic: CCD (Cathode Current Departs). A widespread misconception is that Cathode polarity is always negative.
Cathode rays	Cathode rays are streams of electrons observed in vacuum tubes, i.e. evacuated glass tubes that are equipped with at least two metal electrodes to which a voltage is applied, a cathode or negative electrode and an anode or positive electrode. They were discovered by German scientist Johann Hittorf in 1869 and in 1876 named by Eugen Goldstein kathodenstrahlen . Electrons were first discovered as the constituents of Cathode rays.

Cathode ray tube	The Cathode ray tube is a vacuum tube containing an electron gun (a source of electrons) and a fluorescent screen, with internal or external means to accelerate and deflect the electron beam, used to create images in the form of light emitted from the fluorescent screen. The image may represent electrical waveforms (oscilloscope), pictures (television, computer monitor), radar targets and others. Color Cathode ray tubes have three separate electron guns (shadow mask-type) or electron guns that share some electrodes for all three beams (Sony Trinitron, and licensed versions) The Cathode ray tube uses an evacuated glass envelope which is large, deep, heavy, and relatively fragile.
Time constant	In physics and engineering, the Time constant, usually denoted by the Greek letter τ , is the risetime characterizing the response to a time-varying input of a first-order, linear time-invariant (LTI) system. (That is, a system that can be modeled by a single first order differential equation in time. Examples include the simplest single-stage electrical RC circuits and RL circuits).
Envelope	An Envelope is a packaging product, usually made of flat material such as paper or cardboard, and designed to contain a flat object, which in a postal-service context is usually a letter, card or bills. The traditional type is made from a sheet of paper cut to one of three shapes: the rhombus (also referred to as a lozenge or diamond), the short-arm cross, and the kite. These designs ensure that in the course of Envelope manufacture when the sides of the sheet are folded about a delineated central rectangular area, a rectangular-faced, usually oblong, enclosure is formed with an arrangement of four flaps on the reverse side, which, by virtue of the shapes of sheet traditionally used, is inevitably symmetrical.
Pulse	In medicine, a person`s Pulse is the arterial palpation of a heartbeat. It can be palpated in any place that allows for an artery to be compressed against a bone, such as at the neck (carotid artery), at the wrist (radial artery), behind the knee (popliteal artery), on the inside of the elbow (brachial artery), and near the ankle joint (posterior tibial artery). The Pulse rate can also be measured by measuring the heart beats directly (the apical Pulse).
Impulse	In classical mechanics, an Impulse is defined as the integral of a force with respect to time. When a force is applied to a rigid body it changes the momentum of that body. A small force applied for a long time can produce the same momentum change as a large force applied briefly, because it is the product of the force and the time for which it is applied that is important.

Ramp function

The Ramp function is an elementary unary real function, easily computable as the mean of its independent variable and its absolute value.

This function is applied in engineering (e.g., in the theory of DSP). The name Ramp function can be derived by the look of its graph.

Identity matrix

In linear algebra, the Identity matrix or unit matrix of size n is the n-by-n square matrix with ones on the main diagonal and zeros elsewhere. It is denoted by I_n, or simply by I if the size is immaterial or can be trivially determined by the context. (In some fields, such as quantum mechanics, the Identity matrix is denoted by a boldface one, 1; otherwise it is identical to I).

Analog-to-digital converter

An Analog-to-digital converter is a device which converts continuous signals to discrete digital numbers. The reverse operation is performed by a digital-to-analog converter (DAnalog-to-digital converter).
Typically, an ADC is an electronic device that converts an input analog voltage (or current) to a digital number proportional to the magnitude of the voltage or current.

Equation

An equation is a mathematical statement that asserts the equality of two expressions. equations consist of the expressions that are to be equal on opposite sides of an equal sign, as in

$$x + 3 = 5$$

$$9 - y = 7$$

One use of equations is in mathematical identities, assertions that are true independent of the values of any variables contained within them. For example, for any given value of x it is true that

$$x(x - 1) = x^2 - x\,.$$

However, equations can also be correct for only certain values of the variables.

Automatic control

Automatic control is the research area and theoretical base for mechanization and automation, employing methods from mathematics and engineering. A central concept is that of the system which is to be controlled, such as a rudder, propeller or an entire ballistic missile. The systems studied within Automatic control are mostly the linear systems.

Sensor

A Sensor is a device that measures a physical quantity and converts it into a signal which can be read by an observer or by an instrument. For example, a mercury-in-glass thermometer converts the measured temperature into expansion and contraction of a liquid which can be read on a calibrated glass tube. A thermocouple converts temperature to an output voltage which can be read by a voltmeter.

Telephone

The Telephone is a telecommunications device that transmits and receives sound, most commonly the human voice. It is one of the most common household appliances in the developed world, and has long been considered indispensable to business, industry and government. The word `Telephone` has been adapted to many languages and is widely recognized around the world.

Thermistor

A Thermistor is a type of resistor whose resistance varies with temperature. The word is a portmanteau of thermal and resistor. Thermistors are widely used as inrush current limiters, temperature sensors, self-resetting overcurrent protectors, and self-regulating heating elements.

Causal system

A Causal system (also known as a physical or nonanticipative system) is a system where the output y(t) at some specific instant t_0 only depends on the input x(t) for values of t less than or equal to t_0. Therefore these kinds of systems have outputs and internal states that depend only on the current and previous input values.
The idea that the output of a function at any time depends only on past and present values of input is defined by the property commonly referred to as causality.

Signal

· Signal (information theory), a physical quantity that can carry information.

· Signal processing, the field of techniques used to extract information from Signals

· Cell Signaling, in biology, the system of communication that governs basic cellular activities and coordinates cell actions

· Signal (physiology), an electrochemical communication activity in an organism

· Two theories of `honest Signalling` in different disciplines:

· Signalling theory in evolutionary biology, a theory stating how organisms Signal their state to others with respect to evolutionary survival and reproductive fitness

· Signalling (economics), how market actors can Signal their intentions and values to others in economics

· Signalling game, Signalling theories in evolutionary biology and economics expressed in terms of game theory

· Signal (computing), an event, message, or data structure transmitted between computational processes

· Signal programming, a variety of event-driven programming

· Signals and slots, a Signal-driven software design pattern

· Signals used in various kinds of transport:

· Traffic light

· Railway Signal

· Semaphore

· Beacon

· International maritime Signal flags, using the International Code of Signals

· Signals (military), a name that refers to military communications

· Distress Signal

· Smoke Signal

· Signalling (telecommunications), a part of some communication protocols

· Recognition Signal in human, technical or biological communications

· Signal (band), a Bulgarian rock band

· Signal (American band), an American AOR band

· Signals (album), an album by progressive rock band Rush

· Signal (magazine), a Nazi propaganda magazine published in occupied Europe during World War II

· The Signal, a 2007 horror film

· Signal (toothpaste)

· Signal 1 and Signal 2 radio stations

· Signaal (now Thales Nederland), a Netherlands based defence company

· Signal Systems, a telecommunications company of the McLean Group of Companies

· Signal - The Southeast Electronic Music Festival, an annual music festival held in the United States

· USS Signal, the name of more than one United States Navy ship

· Any implicit, subtle, or disguised message, in situations such as:

· Partnership card games)

· Signals in legal citations

· Kent (game) a card game also known as Signal or Kemp

·

Impulse

In classical mechanics, an Impulse is defined as the integral of a force with respect to time. When a force is applied to a rigid body it changes the momentum of that body. A small force applied for a long time can produce the same momentum change as a large force applied briefly, because it is the product of the force and the time for which it is applied that is important.

Convolution

In mathematics and, in particular, functional analysis, Convolution is a mathematical operation on two functions f and g, producing a third function that is typically viewed as a modified version of one of the original functions. Convolution is similar to cross-correlation. It has applications that include statistics, computer vision, image and signal processing, electrical engineering, and differential equations.

Operational Amplifier

An Operational amplifier, which is often called an op-amp, is a DC-coupled high-gain electronic voltage amplifier with differential inputs and, usually, a single output. Typically the output of the op-amp is controlled either by negative feedback, which largely determines the magnitude of its output voltage gain, or by positive feedback, which facilitates regenerative gain and oscillation. High input impedance at the input terminals (ideally infinite) and low output impedance (ideally zero) are important typical characteristics.

Digital-to-analog Converter

In electronics, a Digital-to-analog converter (DAC or D-to-A) is a device for converting a digital (usually binary) code to an analog signal (current, voltage or electric charge).
An analog-to-digital converter (ADigital-to-analog converter) performs the reverse operation.

Ideally sampled signal.

Transfer function

A Transfer function (also known as the network function) is a mathematical representation, in terms of spatial or temporal frequency, of the relation between the input and output of a (linear time-invariant) system. With optical imaging devices, for example, it is the Fourier transform of the point spread function (hence a function of spatial frequency) i.e. the intensity distribution caused by a point object in the field of view.
The Transfer function is commonly used in the analysis of single-input single-output filters, for instance.

System

In computer science and information science, System could also be a method or an algorithm. Again, an example will illustrate: There are Systems of counting, as with Roman numerals, and various Systems for filing papers, or catalogues, and various library Systems, of which the Dewey Decimal System is an example. This still fits with the definition of components which are connected together (in this case in order to facilitate the flow of information).

Analogue

The analogue switch is an electronic component that behaves in a similar way to a relay, but has no moving parts. The switching element is normally a MOSFET, which is a type of transistor. The control input to the switch is a standard CMOS or TTL logic input, which is shifted by internal circuitry to a suitable voltage for switching the MOSFET. The result is that a logic 0 on the control input causes the MOSFET to have a high resistance, so that the switch is off, and a logic 1 on the input causes the MOSFET to have a low resistance, so that the switch is on.

Causal System

A Causal system (also known as a physical or nonanticipative system) is a system where the output y(t) at some specific instant t_0 only depends on the input x(t) for values of t less than or equal to t_0. Therefore these kinds of systems have outputs and internal states that depend only on the current and previous input values.
The idea that the output of a function at any time depends only on past and present values of input is defined by the property commonly referred to as causality.

Linear differential equation

In mathematics, a Linear differential equation is of the form

$$Ly = f$$

where the differential operator L is a linear operator, y is the unknown function (such as a function of time y(t)), and the right hand side f is a given function of the same nature as y (called the source term). For a function dependent on time we may write the equation more expressively as

$$Ly(t) = f(t)$$

and, even more precisely by bracketing

$$L[y(t)] = f(t)$$

The linear operator L may be considered to be of the form

$$L_n(y) \equiv \frac{d^n y}{dt^n} + A_1(t)\frac{d^{n-1}y}{dt^{n-1}} + \cdots + A_{n-1}(t)\frac{dy}{dt} + A_n(t)y$$

The linearity condition on L rules out operations such as taking the square of the derivative of y; but permits, for example, taking the second derivative of y. It is convenient to rewrite this equation in an operator form

$$L_n(y) \equiv \left[D^n + A_1(t)D^{n-1} + \cdots + A_{n-1}(t)D + A_n(t)\right] y$$

where D is the differential operator d/dt (i.e. Dy = y\` , $D^2y$ = y\`,...

BIBO stability

In electrical engineering, specifically signal processing and control theory, BIBO stability is a form of stability for linear signals and systems that take inputs. BIBO stands for Bounded-Input Bounded-Output. If a system is BIBO stable, then the output will be bounded for every input to the system that is bounded.

Digital

A Digital system is a data technology that uses discrete (discontinuous) values. By contrast, non-Digital (or analog) systems use a continuous range of values to represent information. Although Digital representations are discrete, the information represented can be either discrete, such as numbers, letters or icons, or continuous, such as sounds, images, and other measurements of continuous systems.

Automatic control	Automatic control is the research area and theoretical base for mechanization and automation, employing methods from mathematics and engineering. A central concept is that of the system which is to be controlled, such as a rudder, propeller or an entire ballistic missile. The systems studied within Automatic control are mostly the linear systems.

Chapter 4. FOURIER SERIES

Series	Components of an electrical circuit or electronic circuit can be connected in many different ways. The two simplest of these are called series and parallel and occur very frequently. Components connected in series are connected along a single path, so the same current flows through all of the components.
Signal	· Signal (information theory), a physical quantity that can carry information. · Signal processing, the field of techniques used to extract information from Signals · Cell Signaling, in biology, the system of communication that governs basic cellular activities and coordinates cell actions · Signal (physiology), an electrochemical communication activity in an organism · Two theories of `honest Signalling` in different disciplines: · Signalling theory in evolutionary biology, a theory stating how organisms Signal their state to others with respect to evolutionary survival and reproductive fitness · Signalling (economics), how market actors can Signal their intentions and values to others in economics · Signalling game, Signalling theories in evolutionary biology and economics expressed in terms of game theory · Signal (computing), an event, message, or data structure transmitted between computational processes · Signal programming, a variety of event-driven programming · Signals and slots, a Signal-driven software design pattern · Signals used in various kinds of transport:

· Traffic light

· Railway Signal

· Semaphore

· Beacon

· International maritime Signal flags, using the International Code of Signals

· Signals (military), a name that refers to military communications

· Distress Signal

· Smoke Signal

· Signalling (telecommunications), a part of some communication protocols

· Recognition Signal in human, technical or biological communications

· Signal (band), a Bulgarian rock band

· Signal (American band), an American AOR band

· Signals (album), an album by progressive rock band Rush

· Signal (magazine), a Nazi propaganda magazine published in occupied Europe during World War II

· The Signal, a 2007 horror film

· Signal (toothpaste)

· Signal 1 and Signal 2 radio stations

· Signaal (now Thales Nederland), a Netherlands based defence company

· Signal Systems, a telecommunications company of the McLean Group of Companies

· Signal - The Southeast Electronic Music Festival, an annual music festival held in the United States

· USS Signal, the name of more than one United States Navy ship

· Any implicit, subtle, or disguised message, in situations such as:

· Partnership card games)

· Signals in legal citations

· Kent (game) a card game also known as Signal or Kemp

·

Harmonic

In acoustics and telecommunication, a Harmonic of a wave is a component frequency of the signal that is an integer multiple of the fundamental frequency. For example, if the fundamental frequency is f, the Harmonics have frequencies f, 2f, 3f, 4f, etc. The Harmonics have the property that they are all periodic at the fundamental frequency, therefore the sum of Harmonics is also periodic at that frequency.

Coefficient

In mathematics, a Coefficient is a part of a monomial. Given any divisor of the monomial, the Coefficient is the quotient of the monomial by the divisor. Thus the monomial is the product of the Coefficient and the divisor.

Oscillators

Oscillators inherently produce high levels of phase noise. That noise increases at frequencies close to the oscillation frequency or its harmonics. With the noise being close to the oscillation frequency, it cannot be removed by filtering without also removing the oscillation signal.

Pulse

In medicine, a person`s Pulse is the arterial palpation of a heartbeat. It can be palpated in any place that allows for an artery to be compressed against a bone, such as at the neck (carotid artery), at the wrist (radial artery), behind the knee (popliteal artery), on the inside of the elbow (brachial artery), and near the ankle joint (posterior tibial artery). The Pulse rate can also be measured by measuring the heart beats directly (the apical Pulse).

Sinc function

In mathematics, the Sinc function, denoted by sinc(x) and sometimes as Sa(x), has two definitions. In digital signal processing and information theory, the normalized Sinc function is commonly defined by

$$\operatorname{sinc}(x) = \frac{\sin(\pi x)}{\pi x}.$$

In mathematics, the historical unnormalized Sinc function is defined by

$$\operatorname{sinc}(x) = \frac{\sin(x)}{x}.$$

In both cases, the value of the function at the removable singularity at zero, usually calculated by l`Hôpital`s rule, is sometimes specified explicitly as the limit value 1. The Sinc function is analytic everywhere.

Envelope

An Envelope is a packaging product, usually made of flat material such as paper or cardboard, and designed to contain a flat object, which in a postal-service context is usually a letter, card or bills. The traditional type is made from a sheet of paper cut to one of three shapes: the rhombus (also referred to as a lozenge or diamond), the short-arm cross, and the kite. These designs ensure that in the course of Envelope manufacture when the sides of the sheet are folded about a delineated central rectangular area, a rectangular-faced, usually oblong, enclosure is formed with an arrangement of four flaps on the reverse side, which, by virtue of the shapes of sheet traditionally used, is inevitably symmetrical.

Dirichlet conditions

In mathematics, the Dirichlet conditions are sufficient conditions for a real-valued, periodic function f(x) to be equal the sum of its Fourier series at each point where f is continuous. Moreover, the behavior of the Fourier series at points of discontinuity is determined as well. These conditions are named after Johann Peter Gustav Lejeune Dirichlet.

The conditions are:

· f(x) must have a finite number of extrema in any given interval

· f(x) must have a finite number of discontinuities in any given interval

· f(x) must be absolutely integrable over a period.

· f(x) must be bounded

We state Dirichlet`s theorem assuming f is a periodic function of period 2π with Fourier series expansion

$$f(x) \sim \sum_{n=-\infty}^{\infty} a_n e^{inx},$$

where

$$a_n = \frac{1}{2\pi}\int_{-\pi}^{\pi} f(x)e^{-inx}\,dx.$$

The analogous statement holds irrespective of what the period of f is, or which version of the Fourier expansion is chosen .

Dirichlet`s theorem: If f satisfies Dirichlet conditions, then for all x, we have that the series obtained by plugging x into the Fourier series is convergent, and is given by

$$\sum_{n=-\infty}^{\infty} a_n e^{inx} = \frac{1}{2}(f(x+) + f(x-)),$$

where the notation

$$f(x+) = \lim_{y \to x^+} f(y)$$
$$f(x-) = \lim_{y \to x^-} f(y)$$

	denotes the right/left limits of f.
Gibbs phenomenon	In mathematics, the Gibbs phenomenon is the peculiar manner in which the Fourier series of a piecewise continuously differentiable periodic function behaves at a jump discontinuity: the nth partial sum of the Fourier series has large oscillations near the jump, which might increase the maximum of the partial sum above that of the function itself. The overshoot does not die out as the frequency increases, but approaches a finite limit. These are one cause of ringing artifacts in signal processing.
System	In computer science and information science, System could also be a method or an algorithm. Again, an example will illustrate: There are Systems of counting, as with Roman numerals, and various Systems for filing papers, or catalogues, and various library Systems, of which the Dewey Decimal System is an example. This still fits with the definition of components which are connected together (in this case in order to facilitate the flow of information).
System analysis	System analysis is the branch of electrical engineering that characterizes electrical systems and their properties. Although many of the methods of System analysis can be applied to non-electrical systems, it is a subject often studied by electrical engineers because it has direct relevance to many other areas of their discipline, most notably signal processing and communication systems. A system is characterized by how it responds to input signals.
Spectrum analyzer	A Spectrum analyzer or spectral analyzer is a device used to examine the spectral composition of some electrical, acoustic, or optical waveform. It may also measure the power spectrum. There are analog and digital Spectrum analyzers: · An analog Spectrum analyzer uses either a variable band-pass filter whose mid-frequency is automatically tuned (shifted, swept) through the range of frequencies of which the spectrum is to be measured or a superheterodyne receiver where the local oscillator is swept through a range of frequencies.

Chapter 5. THE FOURIER TRANSFORM

Definition	A definition is a formal passage describing the meaning of a term (a word or phrase). The term to be defined is the definiendum . A term may have many different senses or meanings.
Series	Components of an electrical circuit or electronic circuit can be connected in many different ways. The two simplest of these are called series and parallel and occur very frequently. Components connected in series are connected along a single path, so the same current flows through all of the components.
Sinc function	In mathematics, the Sinc function, denoted by sinc(x) and sometimes as Sa(x), has two definitions. In digital signal processing and information theory, the normalized Sinc function is commonly defined by $\operatorname{sinc}(x) = \frac{\sin(\pi x)}{\pi x}.$ In mathematics, the historical unnormalized Sinc function is defined by $\operatorname{sinc}(x) = \frac{\sin(x)}{x}.$ In both cases, the value of the function at the removable singularity at zero, usually calculated by l`Hôpital`s rule, is sometimes specified explicitly as the limit value 1. The Sinc function is analytic everywhere.
Dirichlet conditions	In mathematics, the Dirichlet conditions are sufficient conditions for a real-valued, periodic function f(x) to be equal the sum of its Fourier series at each point where f is continuous. Moreover, the behavior of the Fourier series at points of discontinuity is determined as well. These conditions are named after Johann Peter Gustav Lejeune Dirichlet. The conditions are: · f(x) must have a finite number of extrema in any given interval · f(x) must have a finite number of discontinuities in any given interval

· f(x) must be absolutely integrable over a period.

· f(x) must be bounded

We state Dirichlet`s theorem assuming f is a periodic function of period 2π with Fourier series expansion

$$f(x) \sim \sum_{n=-\infty}^{\infty} a_n e^{inx},$$

where

$$a_n = \frac{1}{2\pi}\int_{-\pi}^{\pi} f(x)e^{-inx}\,dx.$$

The analogous statement holds irrespective of what the period of f is, or which version of the Fourier expansion is chosen .

Dirichlet`s theorem: If f satisfies Dirichlet conditions, then for all x, we have that the series obtained by plugging x into the Fourier series is convergent, and is given by

$$\sum_{n=-\infty}^{\infty} a_n e^{inx} = \frac{1}{2}(f(x+) + f(x-)),$$

where the notation

$$f(x+) = \lim_{y\to x^+} f(y)$$

$$f(x-) = \lim_{y \to x^-} f(y)$$

denotes the right/left limits of f.

Pulse

In medicine, a person`s Pulse is the arterial palpation of a heartbeat. It can be palpated in any place that allows for an artery to be compressed against a bone, such as at the neck (carotid artery), at the wrist (radial artery), behind the knee (popliteal artery), on the inside of the elbow (brachial artery), and near the ankle joint (posterior tibial artery). The Pulse rate can also be measured by measuring the heart beats directly (the apical Pulse).

Energy

In physics, Energy is a scalar physical quantity that describes the amount of work that can be performed by a force, an attribute of objects and systems that is subject to a conservation law. Different forms of Energy include kinetic, potential, thermal, gravitational, sound, light, elastic, and electromagnetic Energy. The forms of Energy are often named after a related force.

Equation

An equation is a mathematical statement that asserts the equality of two expressions. equations consist of the expressions that are to be equal on opposite sides of an equal sign, as in

$$x + 3 = 5$$

$$9 - y = 7$$

One use of equations is in mathematical identities, assertions that are true independent of the values of any variables contained within them. For example, for any given value of x it is true that

$$x(x - 1) = x^2 - x\ .$$

However, equations can also be correct for only certain values of the variables.

Signal

· Signal (information theory), a physical quantity that can carry information.

· Signal processing, the field of techniques used to extract information from Signals

· Cell Signaling, in biology, the system of communication that governs basic cellular activities and coordinates cell actions

· Signal (physiology), an electrochemical communication activity in an organism

· Two theories of `honest Signalling` in different disciplines:

· Signalling theory in evolutionary biology, a theory stating how organisms Signal their state to others with respect to evolutionary survival and reproductive fitness

· Signalling (economics), how market actors can Signal their intentions and values to others in economics

· Signalling game, Signalling theories in evolutionary biology and economics expressed in terms of game theory

· Signal (computing), an event, message, or data structure transmitted between computational processes

· Signal programming, a variety of event-driven programming

· Signals and slots, a Signal-driven software design pattern

· Signals used in various kinds of transport:

· Traffic light

· Railway Signal

· Semaphore

· Beacon

· International maritime Signal flags, using the International Code of Signals

· Signals (military), a name that refers to military communications

· Distress Signal

· Smoke Signal

· Signalling (telecommunications), a part of some communication protocols

· Recognition Signal in human, technical or biological communications

· Signal (band), a Bulgarian rock band

· Signal (American band), an American AOR band

· Signals (album), an album by progressive rock band Rush

· Signal (magazine), a Nazi propaganda magazine published in occupied Europe during World War II

· The Signal, a 2007 horror film

· Signal (toothpaste)

· Signal 1 and Signal 2 radio stations

· Signaal (now Thales Nederland), a Netherlands based defence company

· Signal Systems, a telecommunications company of the McLean Group of Companies

· Signal - The Southeast Electronic Music Festival, an annual music festival held in the United States

· USS Signal, the name of more than one United States Navy ship

· Any implicit, subtle, or disguised message, in situations such as:

· Partnership card games)

· Signals in legal citations

· Kent (game) a card game also known as Signal or Kemp

·

Impulse

In classical mechanics, an Impulse is defined as the integral of a force with respect to time. When a force is applied to a rigid body it changes the momentum of that body. A small force applied for a long time can produce the same momentum change as a large force applied briefly, because it is the product of the force and the time for which it is applied that is important.

Phasors

In physics and engineering, a phase vector (`phasor`) is a representation of a sine wave whose amplitude (A), phase (θ), and frequency (ω) are time-invariant. It is a subset of a more general concept called analytic representation. Phasors reduce the dependencies on these parameters to three independent factors, thereby simplifying certain kinds of calculations.

Convolution

In mathematics and, in particular, functional analysis, Convolution is a mathematical operation on two functions f and g, producing a third function that is typically viewed as a modified version of one of the original functions. Convolution is similar to cross-correlation. It has applications that include statistics, computer vision, image and signal processing, electrical engineering, and differential equations.

Multiplication

Multiplication is the mathematical operation of scaling one number by another. It is one of the four basic operations in elementary arithmetic (the others being addition, subtraction and division).

Because the result of scaling by whole numbers can be thought of as consisting of some number of copies of the original, Multiplication by whole numbers is equivalent to repeated addition; for example, 3 multiplied by 4 (often said as `3 times 4`) can be calculated by adding 4 copies of 3 together:

$$3 \times 4 = 3 + 3 + 3 + 3 = 12,$$

$$3 \times 4 = 4 + 4 + 4 = 12.$$

There are differences amongst educationalists which number should normally be considered as the number of copies or whether Multiplication should even be introduced as repeated addition.

Transfer function

A Transfer function (also known as the network function) is a mathematical representation, in terms of spatial or temporal frequency, of the relation between the input and output of a (linear time-invariant) system. With optical imaging devices, for example, it is the Fourier transform of the point spread function (hence a function of spatial frequency) i.e. the intensity distribution caused by a point object in the field of view.
The Transfer function is commonly used in the analysis of single-input single-output filters, for instance.

Operational Amplifier

An Operational amplifier, which is often called an op-amp, is a DC-coupled high-gain electronic voltage amplifier with differential inputs and, usually, a single output. Typically the output of the op-amp is controlled either by negative feedback, which largely determines the magnitude of its output voltage gain, or by positive feedback, which facilitates regenerative gain and oscillation. High input impedance at the input terminals (ideally infinite) and low output impedance (ideally zero) are important typical characteristics.

Analog-to-digital converter

An Analog-to-digital converter is a device which converts continuous signals to discrete digital numbers. The reverse operation is performed by a digital-to-analog converter (DAnalog-to-digital converter).
Typically, an ADC is an electronic device that converts an input analog voltage (or current) to a digital number proportional to the magnitude of the voltage or current.

Nyquist rate

In signal processing, the Nyquist rate is two times the bandwidth of a bandlimited signal or a bandlimited channel. This term is used to mean two different things under two different circumstances:

· as a lower bound for the sample rate for alias-free signal sampling (not to be confused with the Nyquist frequency, which is half the sampling rate of a discrete-time system) and

· as an upper bound for the symbol rate across a bandwidth-limited baseband channel such as a telegraph line or passband channel such as a limited radio frequency band or a frequency division multiplex channel.

	The Nyquist rate is the minimum sampling rate required to avoid aliasing, equal to twice the highest frequency contained within the signal.
Reconstruction	In signal processing, reconstruction usually means the determination of an original continuous signal from a sequence of equally spaced samples For a more practical approach based on band-limited signals, see Whittaker-Shannon interpolation formula.
Aliasing	In signal processing and related disciplines, Aliasing refers to an effect that causes different signals to become indistinguishable when sampled. It also refers to the distortion or artifact that results when the signal reconstructed from samples is different than the original continuous signal. Aliasing example of the A letter in Times New Roman.
System	In computer science and information science, System could also be a method or an algorithm. Again, an example will illustrate: There are Systems of counting, as with Roman numerals, and various Systems for filing papers, or catalogues, and various library Systems, of which the Dewey Decimal System is an example. This still fits with the definition of components which are connected together (in this case in order to facilitate the flow of information).
Bode plot	A Bode plot is a graph of the logarithm of the transfer function of a linear, time-invariant system versus frequency, plotted with a log-frequency axis, to show the system`s frequency response. It is usually a combination of a Bode magnitude plot (usually expressed as dB of gain) and a Bode phase plot (the phase is the imaginary part of the complex logarithm of the complex transfer function). Among his several important contributions to circuit theory and control theory, engineer Hendrik Wade Bode (1905-1982), while working at Bell Labs in the United States in the 1930s, devised a simple but accurate method for graphing gain and phase-shift plots.
Frequency spectrum	The Frequency spectrum of a time-domain signal is a representation of that signal in the frequency domain. The Frequency spectrum can be generated via a Fourier transform of the signal, and the resulting values are usually presented as amplitude and phase, both plotted versus frequency. Any signal that can be represented as an amplitude that varies with time has a corresponding Frequency spectrum.

Passband	In brief, the Passband is the range of frequencies or wavelengths that can pass through a filter without being attenuated. In telecommunications, optics, and acoustics, a Passband is the portion of spectrum, between limiting frequencies (or, in the optical regime, limiting wavelengths), that is transmitted with minimum relative loss or maximum relative gain by a filtering device. There are two main categories of digital communication transmission methods: baseband and Passband.
Step response	The Step response of a system in a given initial state consists of the time evolution of its outputs when its control inputs are Heaviside step functions. In electronic engineering and control theory, Step response is the time behaviour of the outputs of a general system when its inputs change from zero to one in a very short time. The concept can be extended to the abstract mathematical notion of a dynamical system using an evolution parameter.
Stopband	A Stopband is a band of frequencies, between specified limits, in which a circuit, such as a filter or telephone circuit, does not let signals through, or the attenuation is above the required Stopband attenuation level. Depending on application, the required attenuation within the Stopband may typically be a value between 20 and 120 dB higher than the nominal passband attenuation, which often is 0 dB. The lower and upper limiting frequencies, also denoted lower and upper Stopband corner frequencies, are the frequencies where the Stopband and the transition bands meet in a filter specification. The Stopband of a low-pass filter is the frequencies from the Stopband corner frequency (which is slightly higher than the passband 3 dB cut-off frequency) up to the infinite frequency.
Digital	A Digital system is a data technology that uses discrete (discontinuous) values. By contrast, non-Digital (or analog) systems use a continuous range of values to represent information. Although Digital representations are discrete, the information represented can be either discrete, such as numbers, letters or icons, or continuous, such as sounds, images, and other measurements of continuous systems.
Frequency spectrum	The Frequency spectrum of a time-domain signal is a representation of that signal in the frequency domain. The Frequency spectrum can be generated via a Fourier transform of the signal, and the resulting values are usually presented as amplitude and phase, both plotted versus frequency.

Any signal that can be represented as an amplitude that varies with time has a corresponding Frequency spectrum.

System

In computer science and information science, System could also be a method or an algorithm. Again, an example will illustrate: There are Systems of counting, as with Roman numerals, and various Systems for filing papers, or catalogues, and various library Systems, of which the Dewey Decimal System is an example. This still fits with the definition of components which are connected together (in this case in order to facilitate the flow of information).

Low-pass filter

A Low-pass filter is a filter that passes low-frequency signals but attenuates (reduces the amplitude of) signals with frequencies higher than the cutoff frequency. The actual amount of attenuation for each frequency varies from filter to filter. It is sometimes called a high-cut filter, or treble cut filter when used in audio applications.

Equation

An equation is a mathematical statement that asserts the equality of two expressions. equations consist of the expressions that are to be equal on opposite sides of an equal sign, as in

$$x + 3 = 5$$

$$9 - y = 7$$

One use of equations is in mathematical identities, assertions that are true independent of the values of any variables contained within them. For example, for any given value of x it is true that

$$x(x - 1) = x^2 - x\,.$$

However, equations can also be correct for only certain values of the variables.

Butterworth filter

The Butterworth filter is one type of electronic filter design. It is designed to have a frequency response which is as flat as mathematically possible in the passband. Another name for it is maximally flat magnitude filter.

Elliptic filter

An Elliptic filter (, for the given values of ripple (whether the ripple is equalized or not). Alternatively, one may give up the ability to independently adjust the passband and stopband ripple, and instead design a filter which is maximally insensitive to component variations. As the ripple in the stopband approaches zero, the filter becomes a type I Chebyshev filter.

Baseband

In telecommunications and signal processing, Baseband is an adjective that describes signals and systems whose range of frequencies is measured from zero to a maximum bandwidth or highest signal frequency; it is sometimes used as a noun for a band of frequencies starting at zero. It can often be considered as synonym to lowpass, and antonym to passband, bandpass or radio frequency (RF) signal.

· A Baseband bandwidth is equal to the highest frequency of a signal or system, or an upper bound on such frequencies.

Signal

· Signal (information theory), a physical quantity that can carry information.

· Signal processing, the field of techniques used to extract information from Signals

· Cell Signaling, in biology, the system of communication that governs basic cellular activities and coordinates cell actions

· Signal (physiology), an electrochemical communication activity in an organism

· Two theories of `honest Signalling` in different disciplines:

· Signalling theory in evolutionary biology, a theory stating how organisms Signal their state to others with respect to evolutionary survival and reproductive fitness

· Signalling (economics), how market actors can Signal their intentions and values to others in economics

· Signalling game, Signalling theories in evolutionary biology and economics expressed in terms of game theory

· Signal (computing), an event, message, or data structure transmitted between computational processes

· Signal programming, a variety of event-driven programming

· Signals and slots, a Signal-driven software design pattern

· Signals used in various kinds of transport:

· Traffic light

· Railway Signal

· Semaphore

· Beacon

· International maritime Signal flags, using the International Code of Signals

· Signals (military), a name that refers to military communications

· Distress Signal

· Smoke Signal

· Signalling (telecommunications), a part of some communication protocols

· Recognition Signal in human, technical or biological communications

· Signal (band), a Bulgarian rock band

· Signal (American band), an American AOR band

· Signals (album), an album by progressive rock band Rush

· Signal (magazine), a Nazi propaganda magazine published in occupied Europe during World War II

· The Signal, a 2007 horror film

· Signal (toothpaste)

· Signal 1 and Signal 2 radio stations

· Signaal (now Thales Nederland), a Netherlands based defence company

· Signal Systems, a telecommunications company of the McLean Group of Companies

· Signal - The Southeast Electronic Music Festival, an annual music festival held in the United States

· USS Signal, the name of more than one United States Navy ship

· Any implicit, subtle, or disguised message, in situations such as:

· Partnership card games)

· Signals in legal citations

· Kent (game) a card game also known as Signal or Kemp

·

Reconstruction	In signal processing, reconstruction usually means the determination of an original continuous signal from a sequence of equally spaced samples For a more practical approach based on band-limited signals, see Whittaker-Shannon interpolation formula.
Aliasing	In signal processing and related disciplines, Aliasing refers to an effect that causes different signals to become indistinguishable when sampled. It also refers to the distortion or artifact that results when the signal reconstructed from samples is different than the original continuous signal. Aliasing example of the A letter in Times New Roman.

Impulse	In classical mechanics, an Impulse is defined as the integral of a force with respect to time. When a force is applied to a rigid body it changes the momentum of that body. A small force applied for a long time can produce the same momentum change as a large force applied briefly, because it is the product of the force and the time for which it is applied that is important.
Nyquist rate	In signal processing, the Nyquist rate is two times the bandwidth of a bandlimited signal or a bandlimited channel. This term is used to mean two different things under two different circumstances: · as a lower bound for the sample rate for alias-free signal sampling (not to be confused with the Nyquist frequency, which is half the sampling rate of a discrete-time system) and · as an upper bound for the symbol rate across a bandwidth-limited baseband channel such as a telegraph line or passband channel such as a limited radio frequency band or a frequency division multiplex channel. The Nyquist rate is the minimum sampling rate required to avoid aliasing, equal to twice the highest frequency contained within the signal.
Digital-to-analog Converter	In electronics, a Digital-to-analog converter (DAC or D-to-A) is a device for converting a digital (usually binary) code to an analog signal (current, voltage or electric charge). An analog-to-digital converter (ADigital-to-analog converter) performs the reverse operation. Ideally sampled signal.
Delta modulation	Delta modulation (Delta modulation or Δ-modulation) is an analog-to-digital and digital-to-analog signal conversion technique used for transmission of voice information where quality is not of primary importance. Delta modulation is the simplest form of differential pulse-code modulation (DPCM) where the difference between successive samples is encoded into n-bit data streams. In Delta modulation, the transmitted data is reduced to a 1-bit data stream.
Demodulation	Demodulation is the act of extracting the original information-bearing signal from a modulated carrier wave. A demodulator is an electronic circuit used to recover the information content from the modulated carrier wave. These terms are traditionally used in connection with radio receivers, but many other systems use many kinds of demodulators.

Intermediate frequency	In communications and electronic engineering, an Intermediate frequency is a frequency to which a carrier frequency is shifted as an intermediate step in transmission or reception. The Intermediate frequency is created by mixing the carrier signal with a local oscillator signal in a process called heterodyning, resulting in a signal at the difference or beat frequency. Intermediate frequencies are used in superheterodyne radio receivers, in which an incoming signal is shifted to an Intermediate frequency for amplification before final detection is done.
Time-division multiplexing	Time-division multiplexing (TDM) is a type of digital or (rarely) analog multiplexing in which two or more signals or bit streams are transferred apparently simultaneously as sub-channels in one communication channel, but are physically taking turns on the channel. The time domain is divided into several recurrent timeslots of fixed length, one for each sub-channel. A sample byte or data block of sub-channel 1 is transmitted during timeslot 1, sub-channel 2 during timeslot 2, etc.
Telephone	The Telephone is a telecommunications device that transmits and receives sound, most commonly the human voice. It is one of the most common household appliances in the developed world, and has long been considered indispensable to business, industry and government. The word `Telephone` has been adapted to many languages and is widely recognized around the world.
Phasors	In physics and engineering, a phase vector (`phasor`) is a representation of a sine wave whose amplitude (A), phase (θ), and frequency (ω) are time-invariant. It is a subset of a more general concept called analytic representation. Phasors reduce the dependencies on these parameters to three independent factors, thereby simplifying certain kinds of calculations.

BIBO stability

In electrical engineering, specifically signal processing and control theory, BIBO stability is a form of stability for linear signals and systems that take inputs. BIBO stands for Bounded-Input Bounded-Output. If a system is BIBO stable, then the output will be bounded for every input to the system that is bounded.

Signal

· Signal (information theory), a physical quantity that can carry information.

· Signal processing, the field of techniques used to extract information from Signals

· Cell Signaling, in biology, the system of communication that governs basic cellular activities and coordinates cell actions

· Signal (physiology), an electrochemical communication activity in an organism

· Two theories of `honest Signalling` in different disciplines:

· Signalling theory in evolutionary biology, a theory stating how organisms Signal their state to others with respect to evolutionary survival and reproductive fitness

· Signalling (economics), how market actors can Signal their intentions and values to others in economics

· Signalling game, Signalling theories in evolutionary biology and economics expressed in terms of game theory

· Signal (computing), an event, message, or data structure transmitted between computational processes

· Signal programming, a variety of event-driven programming

· Signals and slots, a Signal-driven software design pattern

· Signals used in various kinds of transport:

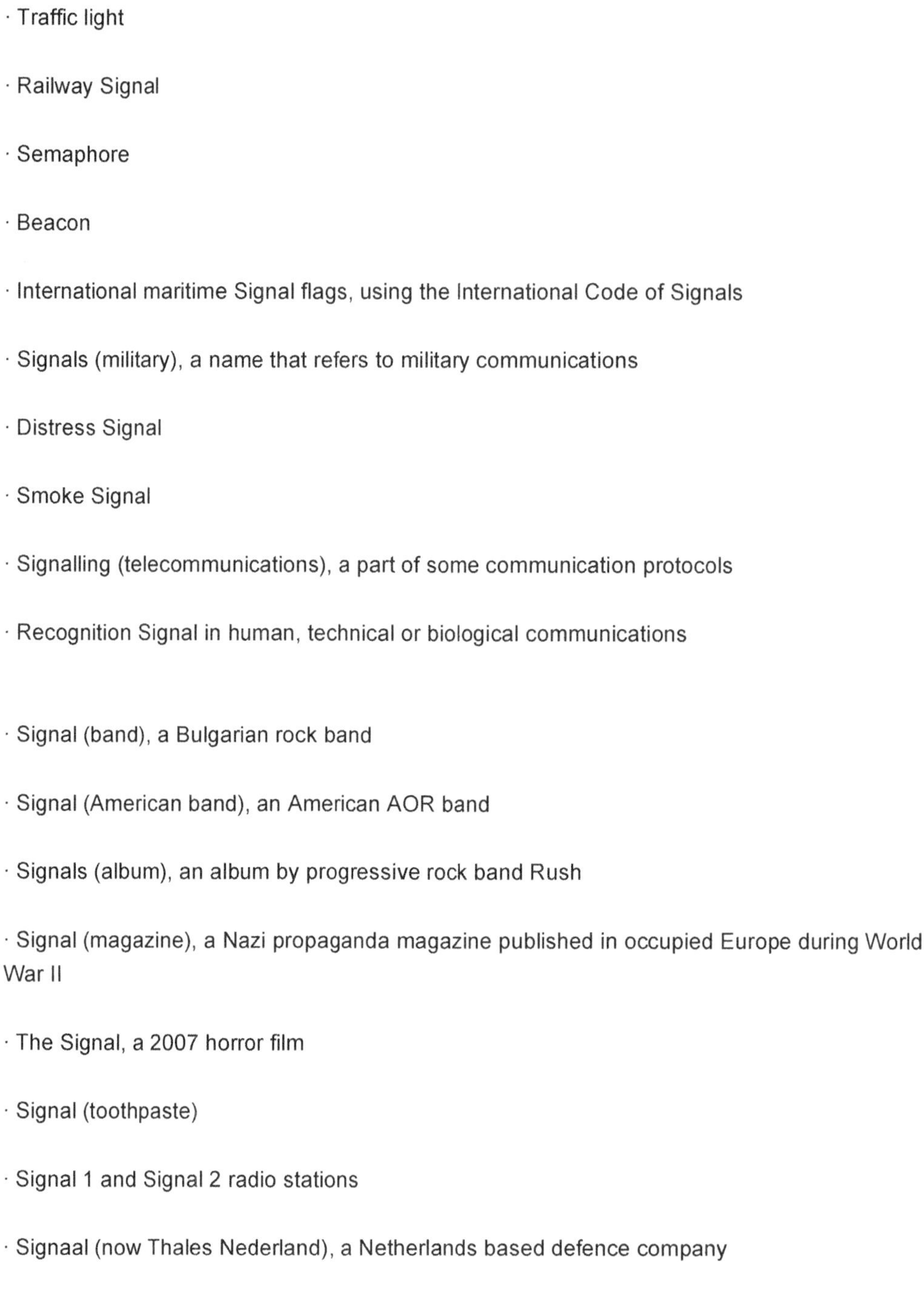

· Traffic light

· Railway Signal

· Semaphore

· Beacon

· International maritime Signal flags, using the International Code of Signals

· Signals (military), a name that refers to military communications

· Distress Signal

· Smoke Signal

· Signalling (telecommunications), a part of some communication protocols

· Recognition Signal in human, technical or biological communications

· Signal (band), a Bulgarian rock band

· Signal (American band), an American AOR band

· Signals (album), an album by progressive rock band Rush

· Signal (magazine), a Nazi propaganda magazine published in occupied Europe during World War II

· The Signal, a 2007 horror film

· Signal (toothpaste)

· Signal 1 and Signal 2 radio stations

· Signaal (now Thales Nederland), a Netherlands based defence company

· Signal Systems, a telecommunications company of the McLean Group of Companies

· Signal - The Southeast Electronic Music Festival, an annual music festival held in the United States

· USS Signal, the name of more than one United States Navy ship

· Any implicit, subtle, or disguised message, in situations such as:

· Partnership card games)

· Signals in legal citations

· Kent (game) a card game also known as Signal or Kemp

.

System In computer science and information science, System could also be a method or an algorithm. Again, an example will illustrate: There are Systems of counting, as with Roman numerals, and various Systems for filing papers, or catalogues, and various library Systems, of which the Dewey Decimal System is an example. This still fits with the definition of components which are connected together (in this case in order to facilitate the flow of information).

Impulse In classical mechanics, an Impulse is defined as the integral of a force with respect to time. When a force is applied to a rigid body it changes the momentum of that body. A small force applied for a long time can produce the same momentum change as a large force applied briefly, because it is the product of the force and the time for which it is applied that is important.

Multiplication Multiplication is the mathematical operation of scaling one number by another. It is one of the four basic operations in elementary arithmetic (the others being addition, subtraction and division).

Because the result of scaling by whole numbers can be thought of as consisting of some number of copies of the original, Multiplication by whole numbers is equivalent to repeated addition; for example, 3 multiplied by 4 (often said as `3 times 4`) can be calculated by adding 4 copies of 3 together:

$$3 \times 4 = 3 + 3 + 3 + 3 = 12,$$
$$3 \times 4 = 4 + 4 + 4 = 12.$$

There are differences amongst educationalists which number should normally be considered as the number of copies or whether Multiplication should even be introduced as repeated addition.

Transfer function

A Transfer function (also known as the network function) is a mathematical representation, in terms of spatial or temporal frequency, of the relation between the input and output of a (linear time-invariant) system. With optical imaging devices, for example, it is the Fourier transform of the point spread function (hence a function of spatial frequency) i.e. the intensity distribution caused by a point object in the field of view.
The Transfer function is commonly used in the analysis of single-input single-output filters, for instance.

Analogue

The analogue switch is an electronic component that behaves in a similar way to a relay, but has no moving parts. The switching element is normally a MOSFET, which is a type of transistor. The control input to the switch is a standard CMOS or TTL logic input, which is shifted by internal circuitry to a suitable voltage for switching the MOSFET. The result is that a logic 0 on the control input causes the MOSFET to have a high resistance, so that the switch is off, and a logic 1 on the input causes the MOSFET to have a low resistance, so that the switch is on.

Causal System

A Causal system (also known as a physical or nonanticipative system) is a system where the output y(t) at some specific instant t_0 only depends on the input x(t) for values of t less than or equal to t_0. Therefore these kinds of systems have outputs and internal states that depend only on the current and previous input values.
The idea that the output of a function at any time depends only on past and present values of input is defined by the property commonly referred to as causality.

Convolution	In mathematics and, in particular, functional analysis, Convolution is a mathematical operation on two functions f and g, producing a third function that is typically viewed as a modified version of one of the original functions. Convolution is similar to cross-correlation. It has applications that include statistics, computer vision, image and signal processing, electrical engineering, and differential equations.
Digital-to-analog Converter	In electronics, a Digital-to-analog converter (DAC or D-to-A) is a device for converting a digital (usually binary) code to an analog signal (current, voltage or electric charge). An analog-to-digital converter (ADigital-to-analog converter) performs the reverse operation. Ideally sampled signal.
Equation	An equation is a mathematical statement that asserts the equality of two expressions. equations consist of the expressions that are to be equal on opposite sides of an equal sign, as in $x + 3 = 5$ $9 - y = 7$ One use of equations is in mathematical identities, assertions that are true independent of the values of any variables contained within them. For example, for any given value of x it is true that $x(x - 1) = x^2 - x\,.$ However, equations can also be correct for only certain values of the variables.
Bode plot	A Bode plot is a graph of the logarithm of the transfer function of a linear, time-invariant system versus frequency, plotted with a log-frequency axis, to show the system`s frequency response. It is usually a combination of a Bode magnitude plot (usually expressed as dB of gain) and a Bode phase plot (the phase is the imaginary part of the complex logarithm of the complex transfer function).

Among his several important contributions to circuit theory and control theory, engineer Hendrik Wade Bode (1905-1982), while working at Bell Labs in the United States in the 1930s, devised a simple but accurate method for graphing gain and phase-shift plots.

Equation

An equation is a mathematical statement that asserts the equality of two expressions. equations consist of the expressions that are to be equal on opposite sides of an equal sign, as in

$$x + 3 = 5$$

$$9 - y = 7$$

One use of equations is in mathematical identities, assertions that are true independent of the values of any variables contained within them. For example, for any given value of x it is true that

$$x(x - 1) = x^2 - x \, .$$

However, equations can also be correct for only certain values of the variables.

System

In computer science and information science, System could also be a method or an algorithm. Again, an example will illustrate: There are Systems of counting, as with Roman numerals, and various Systems for filing papers, or catalogues, and various library Systems, of which the Dewey Decimal System is an example. This still fits with the definition of components which are connected together (in this case in order to facilitate the flow of information).

Signal

· Signal (information theory), a physical quantity that can carry information.

· Signal processing, the field of techniques used to extract information from Signals

· Cell Signaling, in biology, the system of communication that governs basic cellular activities and coordinates cell actions

· Signal (physiology), an electrochemical communication activity in an organism

· Two theories of `honest Signalling` in different disciplines:

· Signalling theory in evolutionary biology, a theory stating how organisms Signal their state to others with respect to evolutionary survival and reproductive fitness

· Signalling (economics), how market actors can Signal their intentions and values to others in economics

· Signalling game, Signalling theories in evolutionary biology and economics expressed in terms of game theory

· Signal (computing), an event, message, or data structure transmitted between computational processes

· Signal programming, a variety of event-driven programming

· Signals and slots, a Signal-driven software design pattern

· Signals used in various kinds of transport:

· Traffic light

· Railway Signal

· Semaphore

· Beacon

· International maritime Signal flags, using the International Code of Signals

· Signals (military), a name that refers to military communications

· Distress Signal

· Smoke Signal

· Signalling (telecommunications), a part of some communication protocols

· Recognition Signal in human, technical or biological communications

· Signal (band), a Bulgarian rock band

· Signal (American band), an American AOR band

· Signals (album), an album by progressive rock band Rush

· Signal (magazine), a Nazi propaganda magazine published in occupied Europe during World War II

· The Signal, a 2007 horror film

· Signal (toothpaste)

· Signal 1 and Signal 2 radio stations

· Signaal (now Thales Nederland), a Netherlands based defence company

· Signal Systems, a telecommunications company of the McLean Group of Companies

· Signal - The Southeast Electronic Music Festival, an annual music festival held in the United States

· USS Signal, the name of more than one United States Navy ship

· Any implicit, subtle, or disguised message, in situations such as:

· Partnership card games)

· Signals in legal citations

· Kent (game) a card game also known as Signal or Kemp

·

Analogue	The analogue switch is an electronic component that behaves in a similar way to a relay, but has no moving parts. The switching element is normally a MOSFET, which is a type of transistor. The control input to the switch is a standard CMOS or TTL logic input, which is shifted by internal circuitry to a suitable voltage for switching the MOSFET. The result is that a logic 0 on the control input causes the MOSFET to have a high resistance, so that the switch is off, and a logic 1 on the input causes the MOSFET to have a low resistance, so that the switch is on.
Series	Components of an electrical circuit or electronic circuit can be connected in many different ways. The two simplest of these are called series and parallel and occur very frequently. Components connected in series are connected along a single path, so the same current flows through all of the components.
Transfer function	A Transfer function (also known as the network function) is a mathematical representation, in terms of spatial or temporal frequency, of the relation between the input and output of a (linear time-invariant) system. With optical imaging devices, for example, it is the Fourier transform of the point spread function (hence a function of spatial frequency) i.e. the intensity distribution caused by a point object in the field of view. The Transfer function is commonly used in the analysis of single-input single-output filters, for instance.

Analog-to-digital converter	An Analog-to-digital converter is a device which converts continuous signals to discrete digital numbers. The reverse operation is performed by a digital-to-analog converter (DAnalog-to-digital converter). Typically, an ADC is an electronic device that converts an input analog voltage (or current) to a digital number proportional to the magnitude of the voltage or current.
Digital	A Digital system is a data technology that uses discrete (discontinuous) values. By contrast, non-Digital (or analog) systems use a continuous range of values to represent information. Although Digital representations are discrete, the information represented can be either discrete, such as numbers, letters or icons, or continuous, such as sounds, images, and other measurements of continuous systems.
Digital signal	The term Digital signal is used to refer to more than one concept. It can refer to discrete-time signals that have a discrete number of levels, for example a sampled and quantified analog signal, or to the continuous-time waveform signals in a digital system, representing a bit-stream. In the first case, a signal that is generated by means of a digital modulation method which is considered as converted to an analog signal, while it is considered as a Digital signal in the second case.
Digital signal processing	Digital signal processing is concerned with the representation of the signals by a sequence of numbers or symbols and the processing of these signals. Digital signal processing and analog signal processing are subfields of signal processing. Digital signal processing includes subfields like: audio and speech signal processing, sonar and radar signal processing, sensor array processing, spectral estimation, statistical signal processing, digital image processing, signal processing for communications, biomedical signal processing, seismic data processing, etc.
Signal	· Signal (information theory), a physical quantity that can carry information. · Signal processing, the field of techniques used to extract information from Signals · Cell Signaling, in biology, the system of communication that governs basic cellular activities and coordinates cell actions · Signal (physiology), an electrochemical communication activity in an organism · Two theories of `honest Signalling` in different disciplines:

· Signalling theory in evolutionary biology, a theory stating how organisms Signal their state to others with respect to evolutionary survival and reproductive fitness

· Signalling (economics), how market actors can Signal their intentions and values to others in economics

· Signalling game, Signalling theories in evolutionary biology and economics expressed in terms of game theory

· Signal (computing), an event, message, or data structure transmitted between computational processes

· Signal programming, a variety of event-driven programming

· Signals and slots, a Signal-driven software design pattern

· Signals used in various kinds of transport:

· Traffic light

· Railway Signal

· Semaphore

· Beacon

· International maritime Signal flags, using the International Code of Signals

· Signals (military), a name that refers to military communications

· Distress Signal

· Smoke Signal

· Signalling (telecommunications), a part of some communication protocols

· Recognition Signal in human, technical or biological communications

· Signal (band), a Bulgarian rock band

· Signal (American band), an American AOR band

· Signals (album), an album by progressive rock band Rush

· Signal (magazine), a Nazi propaganda magazine published in occupied Europe during World War II

· The Signal, a 2007 horror film

· Signal (toothpaste)

· Signal 1 and Signal 2 radio stations

· Signaal (now Thales Nederland), a Netherlands based defence company

· Signal Systems, a telecommunications company of the McLean Group of Companies

· Signal - The Southeast Electronic Music Festival, an annual music festival held in the United States

· USS Signal, the name of more than one United States Navy ship

· Any implicit, subtle, or disguised message, in situations such as:

· Partnership card games)

· Signals in legal citations

· Kent (game) a card game also known as Signal or Kemp

·

System	In computer science and information science, System could also be a method or an algorithm. Again, an example will illustrate: There are Systems of counting, as with Roman numerals, and various Systems for filing papers, or catalogues, and various library Systems, of which the Dewey Decimal System is an example. This still fits with the definition of components which are connected together (in this case in order to facilitate the flow of information).
Analogue	The analogue switch is an electronic component that behaves in a similar way to a relay, but has no moving parts. The switching element is normally a MOSFET, which is a type of transistor. The control input to the switch is a standard CMOS or TTL logic input, which is shifted by internal circuitry to a suitable voltage for switching the MOSFET. The result is that a logic 0 on the control input causes the MOSFET to have a high resistance, so that the switch is off, and a logic 1 on the input causes the MOSFET to have a low resistance, so that the switch is on.
Discrete time	Discrete time is non-continuous time. Sampling at non-continuous times results in discrete-time samples. For example, a newspaper may report the price of crude oil once every 24 hours.
Equation	An equation is a mathematical statement that asserts the equality of two expressions. equations consist of the expressions that are to be equal on opposite sides of an equal sign, as in $x + 3 = 5$ $9 - y = 7$ One use of equations is in mathematical identities, assertions that are true independent of the values of any variables contained within them. For example, for any given value of x it is true that $x(x - 1) = x^2 - x$. However, equations can also be correct for only certain values of the variables.

Identity matrix	In linear algebra, the Identity matrix or unit matrix of size n is the n-by-n square matrix with ones on the main diagonal and zeros elsewhere. It is denoted by I_n, or simply by I if the size is immaterial or can be trivially determined by the context. (In some fields, such as quantum mechanics, the Identity matrix is denoted by a boldface one, 1; otherwise it is identical to I).
Impulse	In classical mechanics, an Impulse is defined as the integral of a force with respect to time. When a force is applied to a rigid body it changes the momentum of that body. A small force applied for a long time can produce the same momentum change as a large force applied briefly, because it is the product of the force and the time for which it is applied that is important.
Definition	A definition is a formal passage describing the meaning of a term (a word or phrase). The term to be defined is the definiendum . A term may have many different senses or meanings.
Convolution	In mathematics and, in particular, functional analysis, Convolution is a mathematical operation on two functions f and g, producing a third function that is typically viewed as a modified version of one of the original functions. Convolution is similar to cross-correlation. It has applications that include statistics, computer vision, image and signal processing, electrical engineering, and differential equations.
Digital-to-analog Converter	In electronics, a Digital-to-analog converter (DAC or D-to-A) is a device for converting a digital (usually binary) code to an analog signal (current, voltage or electric charge). An analog-to-digital converter (ADigital-to-analog converter) performs the reverse operation. Ideally sampled signal.
Time constant	In physics and engineering, the Time constant, usually denoted by the Greek letter τ , is the risetime characterizing the response to a time-varying input of a first-order, linear time-invariant (LTI) system. (That is, a system that can be modeled by a single first order differential equation in time. Examples include the simplest single-stage electrical RC circuits and RL circuits).
Causal system	A Causal system (also known as a physical or nonanticipative system) is a system where the output y(t) at some specific instant t_0 only depends on the input x(t) for values of t less than or equal to t_0. Therefore these kinds of systems have outputs and internal states that depend only on the current and previous input values. The idea that the output of a function at any time depends only on past and present values of input is defined by the property commonly referred to as causality.

Chapter 10. DISCRETE-TIME LINEAR TIME-INVARIANT SYSTEMS

System	In computer science and information science, System could also be a method or an algorithm. Again, an example will illustrate: There are Systems of counting, as with Roman numerals, and various Systems for filing papers, or catalogues, and various library Systems, of which the Dewey Decimal System is an example. This still fits with the definition of components which are connected together (in this case in order to facilitate the flow of information).
Analogue	The analogue switch is an electronic component that behaves in a similar way to a relay, but has no moving parts. The switching element is normally a MOSFET, which is a type of transistor. The control input to the switch is a standard CMOS or TTL logic input, which is shifted by internal circuitry to a suitable voltage for switching the MOSFET. The result is that a logic 0 on the control input causes the MOSFET to have a high resistance, so that the switch is off, and a logic 1 on the input causes the MOSFET to have a low resistance, so that the switch is on.
Discrete time	Discrete time is non-continuous time. Sampling at non-continuous times results in discrete-time samples. For example, a newspaper may report the price of crude oil once every 24 hours.
Signal	· Signal (information theory), a physical quantity that can carry information. · Signal processing, the field of techniques used to extract information from Signals · Cell Signaling, in biology, the system of communication that governs basic cellular activities and coordinates cell actions · Signal (physiology), an electrochemical communication activity in an organism · Two theories of `honest Signalling` in different disciplines: · Signalling theory in evolutionary biology, a theory stating how organisms Signal their state to others with respect to evolutionary survival and reproductive fitness · Signalling (economics), how market actors can Signal their intentions and values to others in economics · Signalling game, Signalling theories in evolutionary biology and economics expressed in terms of game theory

· Signal (computing), an event, message, or data structure transmitted between computational processes

· Signal programming, a variety of event-driven programming

· Signals and slots, a Signal-driven software design pattern

· Signals used in various kinds of transport:

· Traffic light

· Railway Signal

· Semaphore

· Beacon

· International maritime Signal flags, using the International Code of Signals

· Signals (military), a name that refers to military communications

· Distress Signal

· Smoke Signal

· Signalling (telecommunications), a part of some communication protocols

· Recognition Signal in human, technical or biological communications

· Signal (band), a Bulgarian rock band

· Signal (American band), an American AOR band

· Signals (album), an album by progressive rock band Rush

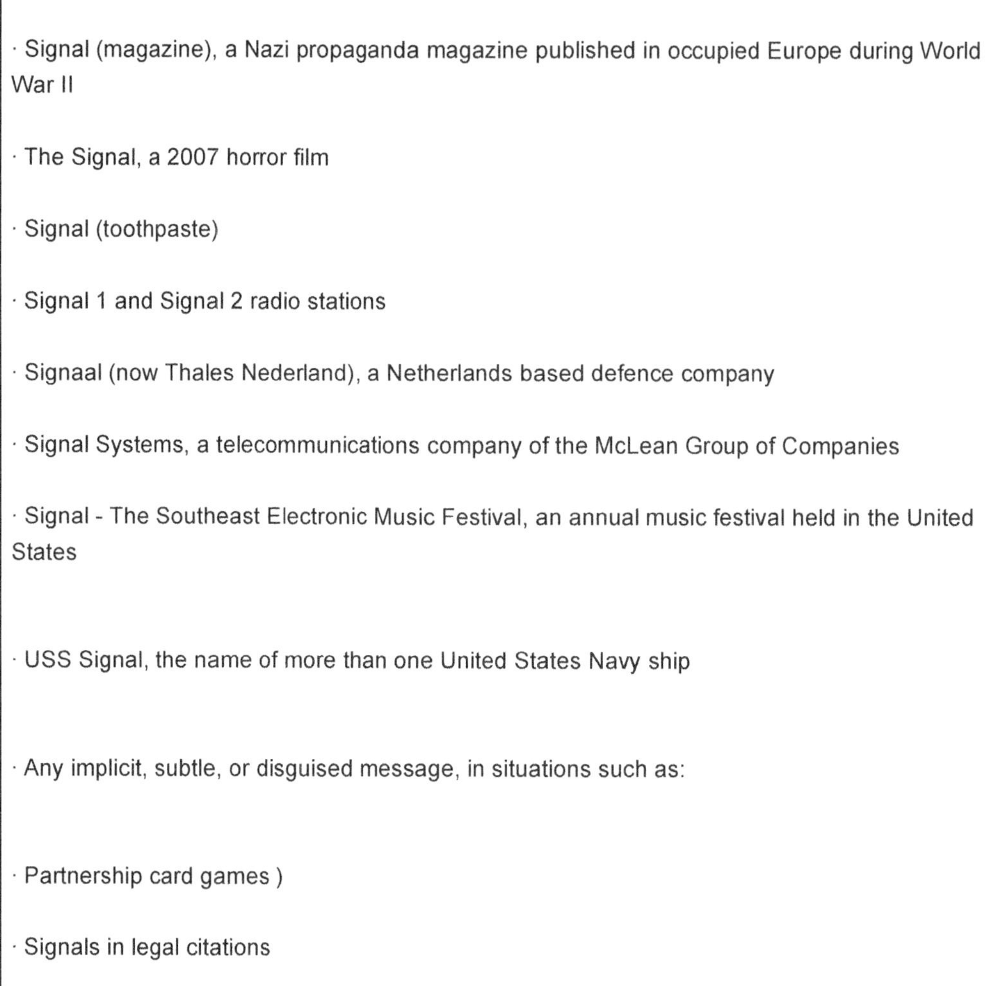

· Signal (magazine), a Nazi propaganda magazine published in occupied Europe during World War II

· The Signal, a 2007 horror film

· Signal (toothpaste)

· Signal 1 and Signal 2 radio stations

· Signaal (now Thales Nederland), a Netherlands based defence company

· Signal Systems, a telecommunications company of the McLean Group of Companies

· Signal - The Southeast Electronic Music Festival, an annual music festival held in the United States

· USS Signal, the name of more than one United States Navy ship

· Any implicit, subtle, or disguised message, in situations such as:

· Partnership card games)

· Signals in legal citations

· Kent (game) a card game also known as Signal or Kemp

·

Impulse

In classical mechanics, an Impulse is defined as the integral of a force with respect to time. When a force is applied to a rigid body it changes the momentum of that body. A small force applied for a long time can produce the same momentum change as a large force applied briefly, because it is the product of the force and the time for which it is applied that is important.

Convolution

In mathematics and, in particular, functional analysis, Convolution is a mathematical operation on two functions f and g, producing a third function that is typically viewed as a modified version of one of the original functions. Convolution is similar to cross-correlation. It has applications that include statistics, computer vision, image and signal processing, electrical engineering, and differential equations.

Finite impulse response

A Finite impulse response (Finite impulse response) filter is a type of a digital filter. The impulse response, the filter`s response to a Kronecker delta input, is finite because it settles to zero in a finite number of sample intervals. This is in contrast to inFinite impulse response (IIR) filters, which have internal feedback and may continue to respond indefinitely.

Infinite impulse response

Infinite impulse response is a property of signal processing systems. Systems with this property are known as Infinite impulse response systems or, when dealing with electronic filter systems, as Infinite impulse response filters. Infinite impulse response systems have an impulse response function that is non-zero over an infinite length of time.

Causal System

A Causal system (also known as a physical or nonanticipative system) is a system where the output y(t) at some specific instant t_0 only depends on the input x(t) for values of t less than or equal to t_0. Therefore these kinds of systems have outputs and internal states that depend only on the current and previous input values.
The idea that the output of a function at any time depends only on past and present values of input is defined by the property commonly referred to as causality.

Digital

A Digital system is a data technology that uses discrete (discontinuous) values. By contrast, non-Digital (or analog) systems use a continuous range of values to represent information. Although Digital representations are discrete, the information represented can be either discrete, such as numbers, letters or icons, or continuous, such as sounds, images, and other measurements of continuous systems.

Digital filter

In electronics, computer science and mathematics, a Digital filter is a system that performs mathematical operations on a sampled, discrete-time signal to reduce or enhance certain aspects of that signal. This is in contrast to the other major type of electronic filter, the analog filter, which is an electronic circuit operating on continuous-time analog signals. An analog signal may be processed by a Digital filter by first being digitized and represented as a sequence of numbers, then manipulated mathematically, and then reconstructed as a new analog signal .

Chapter 10. DISCRETE-TIME LINEAR TIME-INVARIANT SYSTEMS

Equation	An equation is a mathematical statement that asserts the equality of two expressions. equations consist of the expressions that are to be equal on opposite sides of an equal sign, as in $x + 3 = 5$ $9 - y = 7$ One use of equations is in mathematical identities, assertions that are true independent of the values of any variables contained within them. For example, for any given value of x it is true that $x(x-1) = x^2 - x\,.$ However, equations can also be correct for only certain values of the variables.
Coefficient	In mathematics, a Coefficient is a part of a monomial. Given any divisor of the monomial, the Coefficient is the quotient of the monomial by the divisor. Thus the monomial is the product of the Coefficient and the divisor.
Transfer function	A Transfer function (also known as the network function) is a mathematical representation, in terms of spatial or temporal frequency, of the relation between the input and output of a (linear time-invariant) system. With optical imaging devices, for example, it is the Fourier transform of the point spread function (hence a function of spatial frequency) i.e. the intensity distribution caused by a point object in the field of view. The Transfer function is commonly used in the analysis of single-input single-output filters, for instance.
Z-transform	In mathematics and signal processing, the Z-transform converts a discrete time-domain signal, which is a sequence of real or complex numbers, into a complex frequency-domain representation. It can be considered as a discrete equivalent of the Laplace transform. This similarity is explored in the theory of time scale calculus.

Power series

In mathematics, a Power series is an infinite series of the form

$$f(x) = \sum_{n=0}^{\infty} a_n \left(x - c\right)^n = a_0 + a_1(x - c)^1 + a_2(x - c)^2 + a_3(x - c)^3 + \cdots$$

where a_n represents the coefficient of the nth term, c is a constant, and x varies around c (for this reason one sometimes speaks of the series as being centered at c

In many situations c is equal to zero, for instance when considering a Maclaurin series.

Z-transform

In mathematics and signal processing, the Z-transform converts a discrete time-domain signal, which is a sequence of real or complex numbers, into a complex frequency-domain representation.

It can be considered as a discrete equivalent of the Laplace transform. This similarity is explored in the theory of time scale calculus.

Digital

A Digital system is a data technology that uses discrete (discontinuous) values. By contrast, non-Digital (or analog) systems use a continuous range of values to represent information. Although Digital representations are discrete, the information represented can be either discrete, such as numbers, letters or icons, or continuous, such as sounds, images, and other measurements of continuous systems.

Digital filter

In electronics, computer science and mathematics, a Digital filter is a system that performs mathematical operations on a sampled, discrete-time signal to reduce or enhance certain aspects of that signal. This is in contrast to the other major type of electronic filter, the analog filter, which is an electronic circuit operating on continuous-time analog signals. An analog signal may be processed by a Digital filter by first being digitized and represented as a sequence of numbers, then manipulated mathematically, and then reconstructed as a new analog signal .

Frequency scaling

In computer architecture, Frequency scaling is the technique of ramping a processor`s frequency so as to achieve performance gains. Frequency ramping was the dominant force in commodity processor performance increases from the mid-1980s until roughly the end of 2004.

The effect of processor frequency on computer speed can be seen by looking at the equation for computer program runtime:

$$Runtime = \frac{Instructions}{Program} \times \frac{Cycles}{Instruction} \times \frac{Seconds}{Cycles}$$

where instructions per program is the total instructions being executed in a given program, cycles per instruction is a program-dependent, architecture-dependent average value, and seconds per cycles is by definition the inverse of frequency.

Convolution

In mathematics and, in particular, functional analysis, Convolution is a mathematical operation on two functions f and g, producing a third function that is typically viewed as a modified version of one of the original functions. Convolution is similar to cross-correlation. It has applications that include statistics, computer vision, image and signal processing, electrical engineering, and differential equations.

Multiplication

Multiplication is the mathematical operation of scaling one number by another. It is one of the four basic operations in elementary arithmetic (the others being addition, subtraction and division).

Because the result of scaling by whole numbers can be thought of as consisting of some number of copies of the original, Multiplication by whole numbers is equivalent to repeated addition; for example, 3 multiplied by 4 (often said as `3 times 4`) can be calculated by adding 4 copies of 3 together:

$$3 \times 4 = 3 + 3 + 3 + 3 = 12,$$
$$3 \times 4 = 4 + 4 + 4 = 12.$$

There are differences amongst educationalists which number should normally be considered as the number of copies or whether Multiplication should even be introduced as repeated addition.

Chapter 11. THE z-TRANSFORM

Transfer function	A Transfer function (also known as the network function) is a mathematical representation, in terms of spatial or temporal frequency, of the relation between the input and output of a (linear time-invariant) system. With optical imaging devices, for example, it is the Fourier transform of the point spread function (hence a function of spatial frequency) i.e. the intensity distribution caused by a point object in the field of view. The Transfer function is commonly used in the analysis of single-input single-output filters, for instance.
Causal system	A Causal system (also known as a physical or nonanticipative system) is a system where the output y(t) at some specific instant t_0 only depends on the input x(t) for values of t less than or equal to t_0. Therefore these kinds of systems have outputs and internal states that depend only on the current and previous input values. The idea that the output of a function at any time depends only on past and present values of input is defined by the property commonly referred to as causality.
System	In computer science and information science, System could also be a method or an algorithm. Again, an example will illustrate: There are Systems of counting, as with Roman numerals, and various Systems for filing papers, or catalogues, and various library Systems, of which the Dewey Decimal System is an example. This still fits with the definition of components which are connected together (in this case in order to facilitate the flow of information).
Analogue	The analogue switch is an electronic component that behaves in a similar way to a relay, but has no moving parts. The switching element is normally a MOSFET, which is a type of transistor. The control input to the switch is a standard CMOS or TTL logic input, which is shifted by internal circuitry to a suitable voltage for switching the MOSFET. The result is that a logic 0 on the control input causes the MOSFET to have a high resistance, so that the switch is off, and a logic 1 on the input causes the MOSFET to have a low resistance, so that the switch is on.
Equation	An equation is a mathematical statement that asserts the equality of two expressions. equations consist of the expressions that are to be equal on opposite sides of an equal sign, as in $x + 3 = 5$ $9 - y = 7$

One use of equations is in mathematical identities, assertions that are true independent of the values of any variables contained within them. For example, for any given value of x it is true that

$$x(x-1) = x^2 - x\ .$$

However, equations can also be correct for only certain values of the variables.

Discrete frequency

Discrete frequency is defined as the frequency with which the samples of a discrete sinusoid occur. Just as in its continuous-time counterpart , the discrete time signal has a time axis, conventionally denoted by n. The time variable n, however, has a constraint that its continuous time clone does not.

Definition

A definition is a formal passage describing the meaning of a term (a word or phrase). The term to be defined is the definiendum . A term may have many different senses or meanings.

Discrete-time Fourier Transform

In mathematics, the Discrete-time Fourier transform (DTFT) is one of the specific forms of Fourier analysis. As such, it transforms one function into another, which is called the frequency domain representation, or simply the `DTFT`, of the original function (which is often a function in the time-domain). But the DTFT requires an input function that is discrete.

Periodicity

Periodicity is the quality of occurring at regular intervals or periods (in time or space) and can occur in different contexts:

In science:

· Periodicity in time is often specified by its frequency which has the metric units of hertz (the number of periods per second)

· Oscillations, waves and standing waves have crests at periodic intervals of space and/or time

· In physics, period is the number of cycles as a result of time (time/cycle). The amount of time it takes to complete one full revolution. Period is also the inverse of frequency

· In mathematics, a periodic function is a function whose output contains values that repeat periodically

· In mathematics, in group theory, a periodic group is a group in which each element has finite order

· In chemistry, the periodic table is a table which classifies the chemical elements by means of th Periodicity of their chemical properties
Other uses:

· In music theory, Periodicity is described as `predictability gives rise to expectations` .

Circular convolution

For a periodic function $x_T(t)$, with period T, the convolution with another function, h(t), is also periodic, and can be expressed in terms of integration over a finite interval as follows:

$$\begin{aligned} x_T(t) * h(t) &\stackrel{\text{def}}{=} \int_{-\infty}^{\infty} h(\tau)\cdot x_T(t-\tau)\,d\tau \\ &= \int_{t_o}^{t_o+T} h_T(\tau)\cdot x_T(t-\tau)\,d\tau, \end{aligned}$$

where t_o is an arbitrary parameter, and $h_T(t)$ is a periodic extension of h, defined by:

$$h_T(t) \stackrel{\text{def}}{=} \sum_{k=-\infty}^{\infty} h(t-kT) = \sum_{k=-\infty}^{\infty} h(t+kT).$$

When $x_T(t)$ is expressed as the periodic extension of another function, x, this convolution is sometimes referred to as a Circular convolution of functions h and x.

Similarly, for discrete sequences and period N, we can write the Circular convolution of functions h and x as:

∞

	This corresponds to matrix multiplication, and the kernel of the integral transform is a circulant matrix. A case of great practical interest is illustrated in the figure.
Identity matrix	In linear algebra, the Identity matrix or unit matrix of size n is the n-by-n square matrix with ones on the main diagonal and zeros elsewhere. It is denoted by I_n, or simply by I if the size is immaterial or can be trivially determined by the context. (In some fields, such as quantum mechanics, the Identity matrix is denoted by a boldface one, 1; otherwise it is identical to I).
Impulse	In classical mechanics, an Impulse is defined as the integral of a force with respect to time. When a force is applied to a rigid body it changes the momentum of that body. A small force applied for a long time can produce the same momentum change as a large force applied briefly, because it is the product of the force and the time for which it is applied that is important.
Discrete time	Discrete time is non-continuous time. Sampling at non-continuous times results in discrete-time samples. For example, a newspaper may report the price of crude oil once every 24 hours.
Discrete Fourier transform	In mathematics, the Discrete Fourier transform is a specific kind of Fourier transform, used in Fourier analysis. It transforms one function into another, which is called the frequency domain representation, or simply the Discrete Fourier transform, of the original function (which is often a function in the time domain). But the Discrete Fourier transform requires an input function that is discrete and whose non-zero values have a limited (finite) duration.
Analog signal	An Analog or analogue signal is any continuous signal for which the time varying feature (variable) of the signal is a representation of some other time varying quantity, i.e analogous to another time varying signal. It differs from a digital signal in terms of small fluctuations in the signal which are meaningful. Analog is usually thought of in an electrical context; however, mechanical, pneumatic, hydraulic, and other systems may also convey Analog signals.
Fast Fourier transform	A Fast Fourier transform is an efficient algorithm to compute the discrete Fourier transform (DFT) and its inverse A DFT decomposes a sequence of values into components of different frequencies.
Weighting	The process of weighting involves emphasising some aspects of a phenomenon, or of a set of data -- giving them `more weight` in the final effect or result. It is analogous to the practice of adding extra weight to one side of a pair of scales to favour a buyer or seller.

While weighting may be applied to a set of data, for example epidemiological data, it is more commonly applied to measurements of light, heat, sound, gamma radiation, in fact any stimulus that is spread over a spectrum of frequencies.

Distortion

A Distortion is the alteration of the original shape of an object, image, sound, waveform or other form of information or representation. Distortion is usually unwanted. In some fields, Distortion is desirable, such as electric guitar (where Distortion is often induced purposely with the amplifier or an electronic effect to achieve an aggressive sound where desired).

CPSIA information can be obtained at www.ICGtesting.com
Printed in the USA
LVOW03s1457180115

423352LV00001B/8/P